# The Life of Plants

# Plant Parts

Richard & Louise Spilsbury

Heinemann Library
Chicago, Illinois

...ing

...print of Reed Educational & Professional Publishing,
Chicago, Illinois

Customer Service  888-454-2279
Visit our website at www.heinemannlibrary.com

Designed by Macwiz
Illustrated by Jeff Edwards
Originated by Ambassador Litho Ltd
Printed in China by WKT

07 06 05 04
10 9 8 7 6 5 4 3

**Library of Congress Cataloging-in-Publication Data**
Spilsbury, Louise.
  Plant parts / Louise and Richard Spilsbury.
      p. cm. -- (Life of plants)
Includes bibliographical references (p.   ).
Summary: Describes the world of plants and the various parts of specific
plants, such as flower, seed, roots, trunk, and more.
  ISBN 1-4034-0296-5 (HC)    1-4034-0504-2 (PB)
  1. Plants--Juvenile literature. 2.  Botany--Anatomy--Juvenile
literature. [1. Plants. 2. Plant physiology.]  I. Spilsbury, Richard,
1963- II. Title. III. Series.
  QK49 .S735 2002
  580--dc21
                        2001008302

**Acknowledgments**
The authors and publishers are grateful to the following for permission to reproduce copyright material:
pp. 4, 26, 32, 33, 39 Oxford Scientific Films; pp. 5, 7, 8, 9, 10, 11, 12, 14, 16, 18, 22, 25, 28, 29, 31, 35, 38 Holt Studios;
pp. 13, 21 FLPA; p. 15 Bruce Coleman; p. 17 Garden matters; p. 19 Claude Nurisany and Marie Perennou; pp. 24, 34
Science Photo Library; pp. 27, 37 Corbis; p. 30 Andrew Syred; p. 36 Will and Deni McIntyre.

Cover photograph reproduced with permission of Tudor Photography.

Some words are shown in bold, **like this.** You can find out what they mean by looking in the glossary.

# Contents

A plant may be called different things in different countries, so every type of plant has a Latin name that can be recognized anywhere in the world. Latin names are made of two words—the first is the **genus,** or general, group a plant belongs to and the second is its **species,** or specific, name. Latin plant names are given in brackets throughout this book.

# Looking at Plants

There is an astounding variety of plants in the world. They range from tiny, barely visible plants that you might step on without noticing as you walk on a hillside, to huge cactuses in a hot, sandy desert. Yet even though plants all over the world look so incredibly different, most of them are made up of the same basic parts.

## Plants all over

Scientists estimate that there are more than 300,000 **species** of plants in the world, of which around 260,000 are flowering plants. However, no one knows for sure. It is likely that there are even more species of plants out there that have not yet been discovered.

▲ All flowering plants—the most common kind of plants— have **roots**, **stems**, **leaves**, and **flowers**. On some plants, flowers are massive. On others, such as this wood anemone, they are small. Some plants have thin, limp stems, like the wood anemone, while others have very hard, strong stems.

## A job to do

Each part of a plant has a job to do. Plants, like other **organisms**, need food to live, grow, and repair themselves. Animals eat plants, or they eat other animals that eat plants. Plants make their own food using sunlight, water, and **carbon dioxide** in a special process called **photosynthesis.** Some parts of the plant collect these ingredients, and other parts put them together to make food.

Like other organisms, plants also need to **reproduce.** They reproduce so that their species can survive after their death. Plants have special parts that allow them to reproduce when the time is right.

In this book, we'll look at many different plant parts. We'll discover many of the different forms these parts can take and the unique and interesting ways they do their jobs.

## Plant building blocks

All organisms are made up of tiny living parts called **cells.** Cells are so small you can usually see them only with a microscope. Not all plant cells are the same. Different groups of cells form different plant parts, and they help that part do its job. Most plant cells look like minuscule, water-filled balloons, with a tiny, rigid cell wall. Plant cells have these walls; animal cells do not.

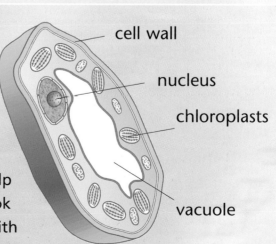

cell wall

nucleus

chloroplasts

vacuole

▲ A magnified cross-section of a typical plant cell.

# Leaves

**Leaves** come in many different sizes, shapes, and arrangements. Leaf shapes are one of the clues that help you identify a plant. For example, the leaves of the banana plant are 6 to 10 feet (2 to 3 meters) long and 1 to 2 feet (30 to 60 centimeters) wide. Aloe leaves are thick with tough skins and jellylike insides.

Leaves can be oval, circular, star-shaped, or heart-shaped. Some leaves, such as spinach, have smooth, rounded edges, but others, such as the European chestnut (*Castanea sativa*), have jagged edges. Most leaves are made up of one piece, like the maple (*Acer*). Others, however, such as those on the ash (*Fraxinus*), are made up of smaller leaves, called leaflets, all joined to a central **stalk**.

## World's longest leaves

The raffia palm (*Raphia farinifera*) has the longest leaves of all palm trees. Its leaves grow up to 65 feet (20 meters) in length. That is about as long as two schoolbuses parked end to end!

◄▲ These photos give you an idea of the huge variety of leaf shapes, sizes, and colors.

## Leaf structure

Even though leaves look very different, they usually have the same parts. The flat part of a leaf is called the **blade.** Leaf blades grow from the **stem** on a stalk, although some grow straight out of the stem. The leaf stalk, called the **petiole,** can bend so that the leaf does not snap off the plant if it is hit by a gust of wind.

The petiole is attached to a **midrib** that runs along the center of the leaf blade. On plants such as the beech (*Fagus*), leaf **veins** branch off from this central midrib. On other plants, such as buffalo grass (*Buchloe dactyloides*), the leaf veins are all lined up from bottom to top. The veins form a network of tubes that connect the leaf to all the other parts of the plant. Water absorbed by a plant's **roots** moves to the leaves through these tubes. Sugary **sap** that carries food from the leaves to other parts of the plant, also travels through them.

A leaf's veins help keep the leaf blade stiff and flat. If you examine a dead leaf rotting under a tree or bush, you can see the leaf "skeleton," or network of veins, very clearly. The leaf skeleton takes longer to break down than the leaf blade because it is made of tougher **cells.**

▼ **Leaves of various species of plants each have unique vein patterns. If you hold a leaf such as this one up to the light, you should be able to see its network of veins.**

blade

vein

midrib

stalk (petiole)

7

# Looking at Leaves

**Leaves** are one of the most important parts of any plant. Leaves help plants get the sunlight and **carbon dioxide** that they need to make their own food, which they do in a process called **photosynthesis.** In photosynthesis, plants use the **energy** from sunlight to combine the carbon dioxide and water, converting them into energy-rich foods, such as sugars.

To get light, many leaves are large and flat because that is the best shape for soaking up a lot of light. Bigger plants need more food to keep them going than smaller plants, so they often have several big leaves or thousands of small ones in order to make enough food.

◄ **The leaves on this banana fan palm are arranged so that they don't overlap one another. That way each leaf gets as much light as possible.**

## Light-catching leaves

Have you ever wondered why leaves grow where they do on plants? They are often positioned high up on a **stem** to receive as much light as possible. Stems and **stalks** hold the leaves out so they can get a share of the light. Sometimes leaves grow alternately—first one side of the stem and then the other, slightly farther up. Some grow in a spiral around the stem or in pairs opposite each other from the same point on the stem.

## Close up

Take a good close look at a leaf. The upper side of the leaf is different from the underside. It often has a waxier, thicker surface than the underside. Most photosynthesis happens in the **cells** of the leaf's upper side, so this is the side that always faces the sun. Warmth from the sun **evaporates** water inside the leaf, so the waxy coat reduces the amount of water the leaf loses.

On the underside of a leaf are many small holes called **stomata**. Stomata are so small that there may be thousands of them in every square inch or centimeter of the leaf's surface. The plant opens its stomata to take in the air it needs for photosynthesis. It can also close them when it is hot to prevent too much water from escaping from the leaf.

## Why are leaves green?

Most leaves are green. This is because they contain a green substance called **chlorophyll**. Chlorophyll is essential for photosynthesis. The leaves of some plants are different colors, such as reddish purple in copper beech (*Fagus purpurea*). Their leaves still contain chlorophyll, but another pigment, the red color, is stronger.

▼ All plants need water to live. The stomata that control the amount of water that plants lose from their leaves can be seen only through a microscope.

# Needles and Spines

Some people find it hard to believe that the long, thin needles of a pine tree and the sharp spines of a cactus plant are really kinds of **leaves.** These leaves grow in special shapes to help the plants they grow on to survive.

Pine trees are **coniferous.** They are able to survive cold weather, when water in the ground may be locked up in ice, because of the form of their leaves. If you look closely at pine needles, you will see that they have a groove on the underside. The leaf's **stomata** are inside the groove. The groove shelters the stomata from cold, dry winds that might **evaporate** water from inside the leaf. A thick, waxy coating on the needles gives them extra protection.

Most cactuses live in hot, dry deserts. Cactus spines are very narrow leaves. Like needles, cactus spines allow very little evaporation. Unlike conifers, which carry out **photosynthesis** in their green needles, cactuses make their food in their thick **stems.** The spines that grow straight out of the stem are often brown. The stems are green because they contain the **chlorophyll** necessary for photosynthesis.

▼ **Cactus spines have many functions. The sharp spines on this cactus help to defend it from animals that may try to eat the watery flesh inside its stem.**

## Kinds of needles

Some needles, such as those of the northern white-cedar (*Thuja occidentalis*), are short, soft, and scale-like. Other needles, such as those of the eastern white pine (*Pinus strobus*), are about four inches (ten centimeters) long and grow in groups of five. Juniper (*Juniperus*) needles grow in groups of three. The leaves of the monkey puzzle tree (*Araucaria araucana*) are tough and triangular and grow in spirals along its branches. The leaves are very sharp to protect the tree's **cones** from being eaten by animals.

▲ These are needles from a conifer.

## Deciduous and coniferous leaves

The broad, flat leaves of **deciduous** trees, such as the European beech (*Fagus sylvatica*), would be damaged by cold winter weather. So these trees drop their leaves in the fall and grow new ones in the spring, when the weather is warmer. In winter, their branches are bare.

Coniferous trees, such as the Scotch pine in the picture at the right, have cones and needlelike leaves. Other familiar conifers are the Norway spruce (*Picea abies*) and the Douglass fir (*Pseudotsuga menziesii*). Their tough, thick, waxy leaves can survive harsh winter weather. Such trees do not need to shed all their leaves in the fall; they shed some old leaves and grow some new leaves throughout the year.

# Buds

In spring, when new **leaves** suddenly appear on **deciduous** trees, where do they come from? The answer is from tiny packages called **buds.** If you look at a tree in winter, the buds are already there, at the ends and along the lengths of the twigs and branches.

A deciduous tree's growth stops or slows down in winter, but its buds are ready to quickly start growing again in spring. Inside the buds, tiny leaves wait, folded neatly in their protective case. When the warmth and light of spring sunshine strikes the buds, they begin to grow and swell until they burst open and the leaves emerge. Once the leaves are out, they can start to make food for the tree.

## Kinds of buds

Different plants have different kinds of buds. The European horsechestnut tree (*Aesculus hippocastanum*) has large, sticky buds. The European ash (*Fraxinus excelsior*) has small buds that are as black as ash from a dead fire. The buds of the lily-flowered magnolia (*Magnolia liliflora*) are covered in soft, thick fur. In winter, when there are no leaves to help you identify a tree, a good look at the buds can help you figure out what it is.

◄ **The buds on a European horsechestnut tree are sticky to defend against insects that might try to eat them.**

# Flower buds

**Flowers** can seem to appear on a plant as if by magic. One day they are not there; the next they are. Flowers are able to appear so swiftly, when the conditions are right, because they are already formed, in miniature, inside buds. Covers called **sepals** protect the fragile, folded young flower **petals** just inside the bud. After the sepals have opened, the petals grow and spread out of a flower bud rapidly. Some kinds of plants, such as tulips (*Tulipa*), produce only one flower, so they grow only one bud. Others, such as apple trees (*Malus*), grow many flowers that burst out of many small, tightly packed buds.

## Hibernating plants?

Most of us have heard of bears hibernating—but plants? **Bulbs** are like underground buds, and they allow certain plants to rest through winter when the cold weather could kill them. Bulbs consist of layers of thick, fleshy leaves surrounding a central bud. A papery skin protects the whole bulb. In the fall, the above-ground plant parts die. The bulb waits underground for the warmth of spring, when it begins to grow new leaves and flowers from its bud. Onions are a well-known kind of bulb.

▲ Tulip flowers burst from their buds. First, the bud swells and grows. Then the petals push their way out of the sepals into the light.

# Flowers

There is an immense variety of **flowers** in the world. Single flowers range in size from just a fraction of an inch, or a few millimeters, to three feet (one meter) across. Some have simple shapes and patterns, while others are incredibly complicated. Flowers may be blue, yellow, white, red, purple, orange, or a combination of colors. Flowers throughout the world may all look very different, but their basic parts are the same.

Flowers grow from **buds.** The swollen end of the flower nearest the **stem** is called a **receptacle.** The other parts of the flower are above the receptacle. On the outside of the flower are the **sepals,** which protect the young, fragile flower parts as they develop. They often look like small green **leaves. Petals** are usually the largest and most colorful part of a flower.

## The reproductive parts

A flower is the center of the plant's **reproductive** system—it is where the **seeds** are made. A flower contains both male and female parts. The female parts are called the **pistils.** At the bottom of the pistils are the **ovaries,** which produce and store the female **sex cells.** The male parts, called **stamens,** are topped by **anthers,** which produce male sex cells, each of which is stored inside a protective **pollen** grain.

anther

pistil

ovary

stamen

▲ This photo of a tulip shows the different parts of a flower.

## How do flowers work?

In order to produce a seed, a male sex cell must join with a female sex cell from the same plant or a different plant of the same **species**. This is called **fertilization**. To do this, pollen must move from an anther to an ovary. This movement, called **pollination,** often happens with the help of insects. This is where the flower's petals and **nectary** come in handy. A nectary is a small sac at the bottom of a petal that contains sweet liquid called **nectar**. Insects visit flowers to feed on the nectar. Pollen  from the anthers brushes onto their bodies. When they land on another flower, the pollen rubs off.

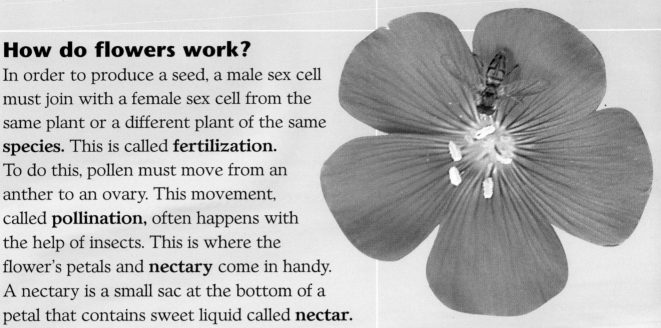

▲ Insects are attracted to flowers by their colorful petals and sweet scent.

If a pollen grain lands at the top of a pistil, the male sex cell inside moves down into the ovary. It joins with the female sex cells, and a seed starts to grow.

## Awesome flowers

The largest single flower in the world is the rafflesia. Once every ten years it produces a foul-smelling flower that grows up to three feet (one meter) wide and weighs fifteen pounds (seven kilograms). However, the titan arum is possibly even more spectacular. It produces a tower of tiny flowers that can grow ten feet (three meters) high!

# Kinds of Flowers

Different **species** of plants have different colors, shapes, and arrangements of **flowers**. The **petals** of bluebells (*Mertensia*) are joined to form a bell. Violets (*Viola*) have petals of different shapes all on one flower. In clematis flowers, it is the **sepals** that are colorful, not the petals. In fuchsias, both the sepals and petals are brightly colored to make doubly sure to draw the attention of insects. Sometimes the sepals and petals look exactly the same. This is common in lilies, for example.

In some flowers, such as sunflowers, dandelions, and daisies, the "petals" are not petals at all; each is part of a tiny flower called a floret. In sunflowers, the center of the flower is made up of florets, too. The whole flowerlike head, called an **inflorescence,** contains hundreds of separate flowers. Each floret is a flower in its own right because each is capable of making a **seed.**

## Flower tricks

Try this trick on your friends. Hold a daisy in your hand and ask them how many flowers you are holding. When they say one, you can surprise them with the fact that actually there are about 250. Pull a daisy to pieces and you will see that it is not just one flower, but many tiny joined flowers, or florets. The yellow central part of the daisy is made up of many tiny tube-shaped florets. On the edge of the daisy, there are different florets, each with a pink petal.

## Flower forms

Each species of plant always grows its flowers in a particular way. The way flowers are arranged on a plant is as important for identifying it as the color or size of its petals.

Some plants, such as the tulip and crocus, grow a single flower out of the end of a single **stem.** Others, such as the scabious (*Scabiosa*), grow many flowers, each one often developing and opening at different times and on different stems. The flowers of some plants grow in spires or spikes. Purple foxglove (*Digitalis purpurea*) flowers grow on little **stalks** coming out of the stem. The flowers at the bottom usually look bigger than the ones at the top because they open first.

Other plants have flowers that grow in groups called **clusters.** Some have many small flowers on separate stalks growing out of the top of the stem. Sometimes these flowers droop, as in Tibetan cowslips (*Primula florindae*). In some plants, such as giant hogweed (*Heracleum mantegazzianum*), many flowers all grow from the same point on the stalk. If the tiny flowers are raised up until they are almost at the same level together, they look a bit like an umbrella. In fact, the name for this shape of flower cluster is an **umbel.**

▲ The umbrellalike shape of an umbellifer, such as this giant hogweed, provides a handy platform for insects.

# What Is a Fruit?

If someone asked you to name three fruits, chances are that you would come up with oranges, apples, pears, or other tasty fruits. But would you think of the "helicopters," sometimes called keys, on sycamores or maple trees, pea **pods,** or cucumbers? The word *fruit* actually means the part of a flowering plant in which **seeds** develop.

Different **species** of fruits can look very different, but they all share two important jobs—to protect developing seeds and to help move them away from the plant. Seeds start to develop if **pollination** has occurred. When a plant has finished flowering, the **petals** on its **flowers** die and fall off. The hollow part at the bottom of the flower, called the **ovary,** starts to swell. The ovary is usually the part that becomes a fruit.

◄ After a tomato flower's petals fall off, its ovary swells into a round, green fruit. The tomato turns red when the seeds inside are ready.

## Apples are not fruits!

Because it is the ovary that holds the seeds, the apple core is the true fruit of an apple tree. The fleshy part of the apple—the sweet, juicy part that we think of as the fruit—is actually the swollen top of the flower stalk, the **receptacle.** This edible part has grown around the real fruit. Other similar "false fruits" include pears and rosehips.

## Kinds of fruits

Some fruits are fleshy. Fleshy fruits such as cherries and apricots contain only one seed surrounded by a hard "stone." Other fleshy fruits, such as melons, cucumbers, and tomatoes, have many seeds scattered throughout their flesh or in a central core. Blackberries and loganberries are compound fruits. Each seed is contained in its own "package" of sweet flesh, and many packages together form the compound fruit. Some plants, such as pumpkins, produce a few giant fruits. Others, such as elder, produce hundreds of small fruits called berries.

Dry fruits include the pods of soybeans, the seed capsules of poppies, and the winged seed cases of plane trees (*Platanus*). The dry fruit of the horsechestnut is hard and spiky to protect the seed inside from being eaten before it drops to the ground.

◄ The dry poppy fruit is like a pepper shaker. As the wind blows, the seeds are shaken out from holes in the top.

### Fruit without seeds

Sometimes you open a fruit such as a grape or banana and find no seeds. Plants rarely lack seeds in the wild because a fruit's most important job is to protect and scatter seeds. But many people don't like to eat fruit with seeds, so growers have developed fruit varieties without seeds.

# Why Plants Have Fruits

One of the things a new young plant needs to grow well is space. It needs room to spread its **roots** and unfold its **leaves.** If a seedling, or young plant, tries to grow when it is too close to other plants, including its parent plant, it will have to compete for the things it needs. Plants cannot move, so how do they ensure that the **seeds** they produce end up somewhere new? The answer is **fruit!** Fruits protect the developing seeds but also have unique ways of dispersing or scattering them.

Fruits come in different shapes, sizes, and forms, depending on the way they spread their seeds. Some plants use the wind, water, or animals to help disperse their seeds. Others have fruits that suddenly split or pop open, catapulting seeds away from the plant.

▲ **The dry fruits of the maple tree have two light, spreading wings. When these fruits fall from the tree, they spin in the air, carrying the seeds away from their parent plant.**

Seeds and fruits that are dispersed by wind, water, or explosion are usually light and fairly small, such as poppy seeds. Maple keys have "wings" so that they can spin away from their parent tree. Dandelion fruits have hairs that, when caught by the wind, act as miniature parachutes, carrying them long distances through the air.

Some fruits, such as those of the alder tree, are dispersed by water. Alder trees grow near rivers, so their fruits drop into the water. Oil in the seed coat around the alder seed helps it float until it lands in a place where it can grow.

Plants that use explosion to disperse their seeds have specially shaped fruits. Peapods twist and dry up when their seeds are fully developed. The twisting makes the two halves of the **pods** burst open, scattering the seeds around.

## Animal assistance

The skins of many fruits are colorful and appealing, and the flesh is sweet and good to eat. This is no accident. When a bird or other animal is attracted to eat a fruit, it often swallows it whole—skin, flesh, and seeds. Later on, the seeds fall to the ground in the animal's droppings. Such seeds usually have extra-tough outer cases so they are not damaged as they pass through the animal's gut. Animals also spit out seeds or drop partly eaten fruit far away from the plant on which the fruit grew.

▲ Orangutans love to eat durians. The strong smell helps lead them to the trees where the fruit grow.

## The forbidden fruit

The durian is a type of very large, spiky fruit that grows in Southeast Asia. The skin of the durian smells like rotting fish, but its custardlike flesh is delicious to eat. In Singapore, people are not allowed to take durians on many taxis, buses, and airplanes because the smell is so unpleasant to most people!

# Seeds

All flowering plants make **seeds.** Seeds are like packages with tiny plants inside. Each seed contains what it needs to start growing into a new plant: an **embryo,** and a supply of plant food. This food is used to keep the embryo alive while it is inside the seed and to give it **energy** to **germinate.** Most seeds also need water, air, and the right temperature to germinate.

Seeds exist in a huge variety of sizes, shapes, and colors. Some are so small that they look like dust or powder. The seeds of some **rain forest** orchids are so tiny that they have no room for food. They depend on **fungi** to supply them with the raw materials necessary for germination. At the other end of the scale, a single coconut seed usually weighs more than five pounds (two kilograms).

Most seeds store their food in special tiny **seed leaves.** The seed leaves also form the first **leaves** that emerge from the seed once it germinates. Some plants, such as corn, have only one seed leaf. Others, such as broad beans, have two.

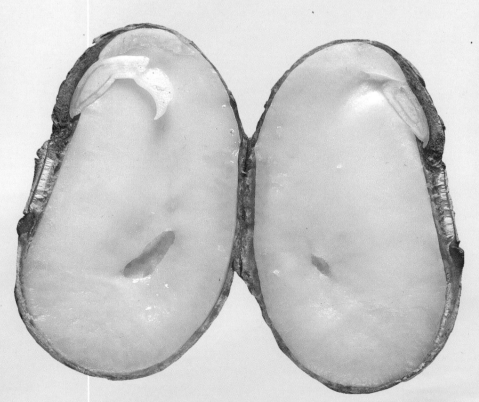

◄ Seeds are like little packages. The seed coat protects the embryo and seed leaves until the conditions are right for it to start to grow into a new plant.

## Seed coats

The outer layer of a seed is called the seed coat. It does three important jobs. It keeps the embryo safe from injury, such as when the seed drops from the plant, for example. It also protects the food-rich seed leaves inside from insects and other animals. The seed coat also keeps the seed from losing water and so prevents it from drying out completely.

## Seeds and germination

Having an efficient seed coat is important because most seeds remain **dormant** rather than germinating right after they leave the parent plant. This rest period prevents seeds from sprouting before the time is right. For example, many seeds remain dormant in cold winter months and germinate in the warmth of spring. Some seeds can survive for years, decades, or even centuries before the conditions are ideal for them to germinate.

▲ When conditions are right, a seed will germinate. A shoot emerges from the seed and grows up toward the light. The first root grows out of the seed and down into the soil.

### Ancient seeds

The seeds of the arctic lupine (*Lupinus arcticus*) can survive for an incredibly long time. Arctic soil is very cold and acts like a freezer, preserving things buried in it. It has been reported that arctic lupine seeds found in a frozen lemming burrow in the Yukon Territory, in Canada, were 10,000 years old. Scientists managed to germinate some of these ancient seeds within 48 hours!

# Nuts

Hundreds of different shrubs and trees produce nuts, but what exactly are nuts? Nuts are actually kinds of **seeds.** They all have a hard or woody outer layer that forms what we call the shell. Inside the shell is the kernel.

The kernel is the seed, and it is also the part that animals, including human beings, like to eat. Kernels are usually rich in **protein** and **fat.** These are the foods that the seed uses to stay alive. It also uses this food supply to grow new **roots** and **shoots** if it gets the chance to **germinate.** Most seeds need water for germination. Water is especially important for nuts because it softens the hard, tough shells so they break apart more easily to release the seeds inside.

shell          kernel

▲ Nut-eating mammals, such as squirrels, have strong teeth to crack open the hard shells of nuts such as those in the photo above.

## Top Seed

The prize for the world's biggest nut goes to the coco-de-mer (*Lodoicea maldivica*). Although its huge seeds can weigh up to 45 pounds (20 kilograms), they can float. The fibers around the shell trap air, and the inside of the nut is hollow. The nut floats from one island to another until it lands somewhere it can germinate.

## Nuts on the move

Just like other seeds, nuts are more likely to germinate if they don't have to compete with their parent plant for light, **nutrients,** and water. So how do nuts move away from their parents? Before winter, animals such as chipmunks and squirrels gather nuts from trees. They eat some and bury the rest, often in open land. When winter comes and food is scarce, chipmunks live off their hidden food supplies. However, the animals never find all the nuts they have hidden. Given the right conditions, the forgotten seeds can grow into new trees in the spring. If a chipmunk happens to bury a nut deep enough, it will be protected by the soil until it is ready to germinate.

## A hard nut to crack

Brazil nuts, as their name suggests, come from Brazil. In the Amazon **rain forest,** local people pick nuts from wild Brazil nut trees (*Bertholletia excelsa*) in the forest. They have to climb very tall trees to get them. The nuts are hidden inside hard wooden cases, which have to be smashed open. Inside, the seeds are protected in smaller, very hard shells. Brazil nuts often have to hang around a long time before conditions are right for them to germinate!

◄ This hard Brazil nut case has been cut in half to show the white nuts inside their shells.

# Cones

Conifer trees do not grow **flowers,** but they still make **seeds.** Conifers have **cones** instead of flowers. Cones are made up of many overlapping scales. There are two kinds of cones—male and female. In most conifers, both kinds of cones grow on the same tree. When people talk about pine cones, they usually are referring to the female cones.

▲ **The female cone on this larch tree (*Larix*) is brown and woody. The male cone (at the top) is smaller and softer.**

The female cones carry the seeds. The male cones are smaller and softer than female cones. They produce and release **pollen,** which is usually scattered by the wind. Female **sex cells** are found on the scales of the large female cones. When pollen lands on the female cones, it **fertilizes** the female cells and seeds start to form.

Conifer seeds lack seed coats, and so they are called naked seeds. They grow between the protective scales of the female cones. The female cones thicken, darken, and harden as the seeds inside their scales develop. Some cones, such as those of the Scotch pine, change from small and green to large, brown, and tough as the seeds develop.

## Kinds of cones

Cones are often smooth and egg-shaped, such as those of cedar trees, but they can be long and thin, such as those of spruces. Some cones, such as those of the Douglass fir and monkey puzzle trees, have prickly scales. Cypress cones are small, knobby, and rounded. The cones of the sugar pine (*Pinus lambertiana*) can grow as long as 25 inches (65 centimeters).

## When the time is right

When the seeds are ready, the cones release them. Most cones take a year to mature, although some pine cones may take three years. Cones release seeds in two ways. Some, like pines, open their scales to release the seeds in dry weather. The seeds of Scotch pines (*Pinus sylvestris*) have a single papery wing that helps them float away from the parent tree. When all the seeds have gone, the cones fall to the ground and gradually rot away.

Other cones, such as those on a cedar tree (*Cedrus*), use a different method. Each cedar scale has two seeds. When the seeds are ripe, the tip of the cone breaks and gradually the scales fall off. They drop to the ground with the seeds attached. That is why it is rare to see a whole cedar cone fallen from a tree. In some other conifers, the seeds are released as the scales fall.

## Fire power!

The Monterey pine (*Pinus radiata*) has cones that refuse to open for anything but the burning heat of a forest fire. The flames destroy other trees, but the Monterey pine is protected by its thick bark. Its cones finally heat up enough to release their seeds, which fall to the ground. Because few other trees survive the fire, the seeds do not have to compete for light and **nutrients,** giving them a good chance of **germinating** and growing into new trees.

# Spores

Ferns, mosses, and liverworts are other kinds of plants that do not grow **flowers,** but unlike **conifer** trees, they do not make **seeds** either. These plants **reproduce** in a different way, by using spores.

Spores are microscopically tiny flecks of living material. When released from the plant, they are able to grow into a clone—that is, a new plant that is identical to the parent plant. Different kinds of plants produce spores of different sizes and shapes, but most spores consist of a single **cell.** Spores are so small that they are very hard to see, but they are grouped together in packets called **sporangia.** You may have seen sporangia before. They are the brown bumps found on the underside of many fern fronds, or **leaves.**

Plants release their spores when they are ripe. Spores have a thick outer coat that protects them against harsh conditions. This outer coat enables them to remain **dormant** for months, or even years, until conditions are right for them to develop into new plants. Most moss spores need damp, shady places. Ferns, mosses, and horsetails produce millions of spores to ensure that at least some have a chance of landing in conditions suitable for growth.

▶ **The brown spots on the underside of this fern frond are the sporangia that produce the plant's spores. They hold the spores until they are ready to be released.**

## Spore containers

Different plants produce sporangia that look very different. Mosses and liverworts are the plants that form velvety green carpets over forest floors, damp walls, or riverside tree trunks. In spring, mosses produce **stalks** with a small oval capsule at the tip. Spores grow inside the capsule.

Horsetails (*Equisetum*) make their spores in structures that resemble "cones." These cones grow at the end of tubelike **stems.** The cones of the great horsetail (*Equisetum telmateia*) are a couple inches, or several centimeters, long on stems about three feet, or one meter, high, surrounded by rings of needle-shaped leaves.

The sporangia of different **species** of fern have different shapes, too. On bracken (*Pteridium*) they are circular, but on hart's tongue fern (*Phyllitis scolopendrium*) they are like stripes. The sporangia of maidenhair ferns (*Adiantum*) are protected behind little flaps along the edges of the frond.

# Stems

Stems come in all shapes and sizes. They range from the flimsy green stem of a primrose to the thick, brown, woody trunk of a tree. In bamboo, the stem is the largest part of the plant. In other plants, such as cabbages, the stem is so short that the cabbage plant appears not to have a stem at all. The stems of many roses are covered in thorns to prevent animals from eating them.

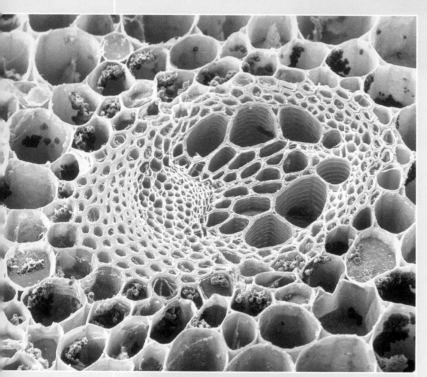

▲ If you look at a cross-section of a stem such as the one above, you can see the tubes that carry water and food to all parts of the plant.

Most stems grow up from the ground. Their job is to hold the **leaves** up to the light so that **photosynthesis** can take place and to hold **flowers** in the best position for **pollination**. A stem's other job is to transport water and food to all parts of the plant. It moves these substances in tubes. One set of tubes, called the **xylem**, carries water and **nutrients** from the **roots**. Other tubes, called **phloem**, carry food made in the leaves to the rest of the plant.

## Going underground

Most stems grow upright but, for protection, some grow partly underground. They may stay underground or grow sideways through the soil. Swollen stems that grow beneath the ground on potato plants are called **tubers**. The potato tuber is the plant's food supply—and the part that people eat. On flowering plants such as irises, such swollen stems are called **rhizomes**.

## Water power

Many plant stems are quite hard and firm. As a result, they can stand up straight and support the weight of the leaves and flowers. Most plant stems, however, are supported by water pressure. The flow of water through the stem helps them stay upright. In dry conditions, a plant does not get enough water to support its stems, and so the plant wilts, or droops.

## Going up!

Some plants have weak, limp stems that cannot support themselves. Runners are horizontal stems that grow along the ground. At intervals along their length, strawberry runners produce flowers, leaves, and roots that form new plants when the runner separates from the parent plant.

Climbers have weak stems, but they manage to grow upward. Some climbers, such as morning glories (*Ipomoea*), coil their wiry stems around other plants' stems to help them reach the light. Tiny roots grow from the stems of ivy plants. These roots work their way into cracks and crevices on walls, firmly anchoring the plant.

Some plants, such as pea plants and clematis, send out thin, curly **tendrils.** The tips of tendrils are like long fingers, and they are sensitive to touch. They grow out randomly and when they touch something, they curl around it.

▼ The stems on a wisteria plant twine around poles or wires for support. Then they become thick or woody.

# Trunks and Branches

Trunks and branches are the **stems** of a tree. Just like the smaller stems of other plants, their job is to hold up **leaves** and **flowers**. The difference is that the tough, thick stems on trees are made of wood. Trunks need to be extra strong because they have to support heavy branches, twigs, and many leaves.

The trunk and branches of a tree are also the places where water, food, and **nutrients** move around. Water and nutrients are taken in from the **roots** and sucked up the trunk and along the branches to all the parts of the tree. At the same time, food is carried away from the leaves to the rest of the tree. This movement of food and water goes on constantly, just underneath the layer of bark that wraps itself around the outside of the tree like a skin.

## The ballooning baobab tree

The trunk of the African baobab tree (*Adansonia digitata*), unlike most other trunks, swells and shrinks throughout the year. These strange-looking trees, which grow up to 80 feet (25 meters) tall, have very small branches compared to their wide, thick trunks. In the parts of Africa where they grow, there are long dry spells. When a rainy spell comes, the trunk of the baobab tree swells with a huge water supply. As this water is used up, the trunk becomes slim again.

◄ Baobab trees have distinctive chubby trunks topped with small branches.

## Inside a trunk

If you look at the cut surface of a log you will see many rings, called **growth rings,** radiating from the center. Each ring represents one year's growth. You can find out how old a tree is by counting the number of growth rings.

You can also see layers of different types of wood under the bark. The middle part of the trunk is called heartwood. It is made up of dead, woody **cells** that have become very hard. The heartwood helps to make the tree trunk strong. Closer to the surface is a very thin layer made up of living cells. This is the sapwood, where water and nutrients are carried up through the **xylem** from the roots. It is also where **phloem** tubes carry food made by the leaves to the rest of the tree. The outermost layer is the bark, which protects the sapwood beneath it.

### Rubber trees?

Next time you bounce a rubber ball, think about the tree it came from. Most natural rubber is made from white latex or **sap** from rubber trees (*Hevea brasiliensis*). Workers make cuts in the bark and collect the oozing latex in cups attached to the tree.

◄ By counting the growth rings on a tree, you can tell how many years old it was when it was cut down.

heartwood

sapwood

bark

# Bark

Bark is like the skin on your body. It is a tough outer layer that covers the main body of the tree—its trunk and branches—and helps protect the parts inside.

▲ The reddish color and curling ends of its peeling outer layer of bark identifies this tree as a paperbark maple.

Different trees have different types of bark that can help you identify the tree **species** you are looking at. The bark of a European white birch (*Betula pendula*) is papery and silvery-white in color. As they grow older, the bark of many **coniferous** trees flakes off in small pieces like those of a jigsaw puzzle. Some trees have distinctive bark patterns, such as the circles on a monkey puzzle tree (*Araucaria Araucana*) or the short stripes on a cherry tree (*Prunus*).

The bark of a tree usually gets thicker as the tree gets older. However, on an old beech tree, the bark is thin because older layers flake and drop off. This does not happen to the bark of a redwood tree (*Sequoia sempervirens*), which can be 12 inches (30 centimeters) thick. The redwood's thick bark helps make the tree fire resistant.

## Cracking and splitting

Bark does not stretch as skin does. It cracks, splits, or peels as a tree grows. Younger trees usually have smooth bark. As the trunk grows thicker, the outer layers of bark on the outside split or peel as newer layers inside push outward.

# The living dead

We all know that trees are living things, but did you know that, apart from the **buds** and **leaves,** the rest of the visible parts of the tree is dead? Bark is made up of two layers. The inner layer that we cannot see is the living, growing part. Every year, this special layer of growing **cells** makes a new ring of bark. The new bark pushes the older bark outward. The old bark is cut off from the nourishing fluids moving through the **xylem** and **phloem,** in the tree, and it dies. Then it becomes part of the visible outer layer of dead bark.

Bark does several very important jobs. It protects the xylem and phloem inside the tree trunk. If a strip of bark all the way around a tree trunk is damaged, the food and water supply to all other parts of the tree is cut off and the whole tree can die.

Bark also protects the tree from damage by insects and **fungi.** If woodland animals, such as deer, chew bark, they leave a hole or wound in the tree. If the wound stays open, insects or fungi damage the live wood underneath. Some trees release a sticky liquid, called **sap,** from their phloem to plug small holes in their bark. Sap can harden and seal the wound.

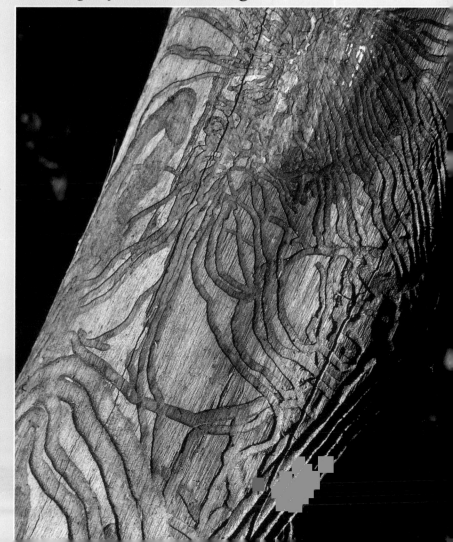

► **Chewing insects may leave burrows under a tree's bark. If large numbers of insects attack wood in this way, the tree can be damaged.**

# Roots

The **roots** of a plant are like the foundation of a building. Underground foundations support buildings and provide a place for pipes and cables that carry water and electricity into the building. Similarly, roots anchor a plant firmly in the soil, holding it in place even in high winds. They also absorb water and **nutrients** that are vital for the plant's survival.

## Soil secrets

People need different kinds of nutrients to stay healthy, such as the **vitamins** found in fruit. Plants need a variety of nutrients, too. They make some by **photosynthesis,** but they get other nutrients they need from the soil. Useful **minerals** come from the rock particles that form part of the soil. Other nutrients come from dead plant and animal matter that has decayed to become part of the soil.

## Kinds of roots

Even though most roots do the same jobs, roots have many different shapes and sizes. The kind of roots a plant has may depend on how large the plant grows and the kind of **habitat** it lives in. For example, in their search for water and nutrients, plants that grow where water is scarce or the soil is poor grow much deeper roots than other plants.

▶ Although it is sad when a storm blows down a giant tree, it's a chance to see how far and wide its roots have spread.

## How do roots work?

As a root grows, smaller roots branch out from it. The larger roots are tougher and have thicker skin. They help to hold the plant in place. The smallest roots are covered with many tiny, thin-skinned hairs. These root hairs suck up water and nutrients from the soil. The water and nutrients go into **xylem** tubes in the larger roots and travel up through the **stem** or trunk to the **leaves**.

▲ Even though roots may not look strong, they are. They have to be strong to push their way through soil that is tightly packed and often rocky. Some roots even have the strength to break apart buildings or rocks. If they grow into cracks in rocks or buildings, the roots can widen the gap further as they grow wider.

## Longest roots

The winner of the record for the longest roots of any plant is the winter rye plant (*Secale cereale*). The winter rye is a kind of grass plant. It can produce more than 385 miles (620 kilometers) of tiny roots curled up into a volume of soil the size of a tin can!

# Kinds of Roots

There are two main kinds of **roots:** fibrous roots and taproots. Plants with many roots, all of which are roughly the same size, are said to have fibrous roots. The many roots spread out in all directions under the ground, but they do not go very deeply into the soil. Plants that have fibrous roots include grasses and many trees. The spread of a tree's roots below ground usually matches the width of its crown, the branches and **leaves** above the trunk.

A plant with a taproot, such as the dandelion, has one root that is bigger than all the rest. This main root grows straight down into the soil. Smaller, thinner roots grow out from the side of it. In some plants with taproots, such as carrots and radishes, the taproot becomes swollen with food made by leaves of the plant. This is the part we eat. The plant stores the food this way so it can use it if conditions change and it is unable to make its own food. Some trees, such as oaks and pines, have taproots.

▶ **This marigold plant grows a taproot that is larger than the other roots, which may grow sideways from the taproot.**

## Remarkable roots

Special **habitats** require special types of roots. In the Amazon **rain forest**, some types of fig trees are more than 130 feet (40 meters) tall, but they grow in shallow soil. To prevent them from toppling over, these trees have special wedge-shaped roots above the surface of the ground. These buttress roots grow up the side of the trunks and act like feet, helping the tree stand firm (see right). In Australian **mangrove** swamps, the mud can be so thick and wet that the mangroves' roots cannot get the air they need. So, the trees produce some roots that grow upward out of the water and into the air.

## Plant parts work together

Each of the plant parts we have looked at in this book is different. The parts look different and they have different jobs to do, but they all work together to form a whole. The roots and leaves supply the plant with the **energy** and **nutrients** it needs. Vessels in the roots, leaves, and **stem** carry water, food, and nutrients throughout the plant. These substances nourish the plant, enabling it to grow and produce other parts, such as **flowers, fruits, cones,** and **spores.** These, in turn, allow the plant to produce **seeds** that may grow into new plants to begin the cycle again. A plant is kind of like a human body. It may be made up of separate parts, but the individual parts need to work together if the plant is to live, grow, and thrive.

# Try It Yourself!

Try these experiments and activities to find out more about some of the plant processes you have learned about in this book.

## Flower power

You can make a **flower** any color you like by using the power in its **stem**.

**You will need:**

- two glasses
- water
- two white flowers (carnations are ideal)
- blue and red food coloring (or ink)

Cut about 2 in. (5 cm) off the bottom of each flower stem. Look at the cut end. Can you see any tubes?

Fill one glass halfway with water and add a few drops of red coloring. Fill the second glass halfway with water and add a few drops of blue coloring.

Put one flower in each glass. Leave them for a day.

What happens? Your white flowers should change color. As the stem takes water from its base to the flower, it also takes up the food coloring, which dyes the **petals** a different color.

## Bark rubbings

Taking bark rubbings is fun and a great way to examine different trees' bark patterns.

**You will need:**

- a pencil
- brown crayons
- large sheets of paper

Hold a sheet of paper against the bark of a tree. You may need someone to hold the paper for you. Take the paper wrapping off the crayon and, holding the crayon flat on its side, rub it over the paper. Try to cover the whole sheet. Use the pencil to write the name of the tree on the back of the paper.

Repeat this process for different **species** of trees.

You can put your rubbings in a scrapbook or use them to make a display. Label each rubbing with the name of the tree it came from.

## Hyacinth happenings

When you grow a hyacinth **bulb** in a glass jar, it is easy to see the roots that grow from it.

**You will need:**

- a hyacinth bulb
- a glass jar with a narrow top (about the width of the bulb)
- water
- a magnifying glass

Fill the jar with enough water to barely cover the bottom of the bulb when the bulb is sitting on the top of the jar. Place the bulb in the jar and put it in a warm, bright spot, such as a windowsill.

When roots begin to grow from the bulb, look closely at them with the magnifying glass. You should be able to see the root hairs growing out from the roots.

These root hairs help the roots cover an even wider area as they reach out to collect water for the plant.

If you continue to care for your plant, keeping the water at the proper level, it should produce a colorful, scented flower after a few weeks.

## Plant parts to eat

Have you ever really thought about which part of a plant you are eating when you munch on **fruit** or vegetables?

**You will need:**

- a notepad
- a pencil

Have an adult take you to a supermarket. Look at all the different fruits and vegetables and try to identify which part of the plant they are. You could organize your findings in a chart. Under the heading "Kind of food," you would write the name of the fruit or vegetable, such as broccoli or celery. Under "Part of plant," you would write *flower* and *stem*.

By the way, did you also know that you should try to eat five servings of fruits or vegetable every day to be really healthy? Which five are your favorites?

# Looking at Plant Parts

## Naming plant parts

This diagram of a plant shows the parts that most plants have in common and what those parts do.

**Flowers** are the plant's **reproductive** parts. They make **seeds**, which may grow into new plants, like the parent plant.

**Buds** hold tiny **leaves** or flowers until the time is right for them to open.

**Fruits** protect the seeds as they grow. They also help spread the seeds away from the parent plant so that the seeds have a better chance of growing.

Seeds are made in the flowers. When they land in a suitable spot, they grow into new plants.

Leaves are the plant's food factories. They collect sunlight, which they use to convert water and **carbon dioxide** into food.

The **stem** supports the other parts of the plant. It holds up the leaves to collect sunlight and holds up the flowers so insects or wind can disperse their **pollen** and seeds.

**Roots** anchor the plant in the ground and absorb water and **nutrients** through root hairs. Root hairs are usually too tiny to see.

## Leaf shapes

### Simple leaves

Garden nasturtium (*Trapaeolum majus*)

English ivy (*Hedera helix*)

Devilwood (*Osmanthus americanus*)

Pacific dogwood (*Cornus nuttallii*)

### Simple straplike leaves

New Zealand flax (*Phormium*)

Tulip (*Tulipa*)

Bamboo (*Arundinaria*)

Iris (*Iris*)

### Toothed simple leaves

Holly (*Ilex*)

White alder (*Alnus rhombifolia*)

Pussy willow (*Salix discolor*)

Chestnut (*Castanea*)

### Lobed simple leaves

Sycamore (*Acer pseudoplatanus*)

Sugar maple (*Acer saccharum*)

Coastal sage scrub oak (*Quercus dumosa*)

Yellow-poplar, or tulip tree (*Liriodendron tulipifera*)

## Needles and spines

Scotch pine (*Pinus sylvestris*)

Redwood (*Sequoia sempervirens*)

Giant sequoia (*Sequoiadendron giganteum*)

Australian pine (*Casuarina equisetifolia*)

## Compound leaves

Ash (*Fraxinus*)

Horse chestnut (*Aesculus*)

Peashrub (*Caragana*)

Coconut palm (*Cocos nucifera*)

# Arrangements of flowers

Some plants bear single flowers that grow at the top of a stem, but most flowers grow in characteristic arrangements. These are listed below with examples for each shape.

## Simple flowers

Carnation (*Dianthus*)

Iris (*Iris*)

Oriental poppy (*Papaver orientale*)

Tulip (*Tulipa*)

Buttercup (*Ranunculus*)

## Spike

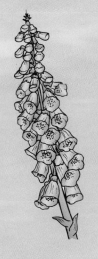

Many flowers grow from a stem, each on their own little **stalk**. The oldest flower is at the bottom, the youngest at the top.

Foxglove (*Digitalis*)

Yucca (*Yucca*)

Orchid (*Orchidaceae*)

Pampas grass (*Cortaderia*)

## Umbel

Umbels have umbrella-like arrangements of flowers. At the end of a stem are flower **stalks** of equal length, each with a single flower or tiny group of flowers. The flowers may be held upright or hang downward.

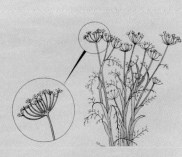

Flowering garlic (*Allium*)

Giant hogweed (*Heracleum mantegazzianum*)

Angelica (*Angelica*)

Leafy spurge (*Euphorbia esula*)

## Composites—flower clusters

This arrangement of flowers looks like a single flower. But the flower head is actually made up of hundreds of tiny flowers packed closely together.

Oxeye daisy (*Chrysanthemum leucanthemum*)

Sunflower (*Helianthus*)

Clover (*Trifolium*)

Edelweiss (*Leontopodium alpinum*)

# Glossary

**algae** group of plantlike organisms that lack leaves, stems, roots, and flowers. Like plants, algae can use sunlight to make their own food in the process of photosynthesis.

**anther** swollen tip of a stamen, the male reproductive parts of a flower. Each anther is made up of pollen sacs, which contain grains of pollen.

**blade** flat part of a leaf

**bud** swelling on a plant stem containing tiny young overlapping leaves or petals and other flower parts that are ready to burst into bloom

**bulb** underground bud protected by layers of thick, fleshy leaves. An onion is a kind of bulb.

**carbon dioxide** gas in the air that plants use for photosynthesis

**cell** building block of living things, so small it can only be seen with a microscope. Some microbes consist of a single cell, but most plants and animals are made up of millions or billions of cells.

**chlorophyll** green substance in plants that is used in photosynthesis. Chlorophyll gives leaves their green color.

**cluster** group of flowers growing together

**cone** form of dry fruit produced by a conifer. Cones are often egg-shaped, and they are made up of many overlapping scales in which seeds grow.

**conifer/coniferous** kind of tree that has cones and needlelike leaves

**deciduous** describes a kind of tree that loses all its leaves in winter. In cool climates, deciduous trees lose all their leaves before winter. In tropical areas, deciduous trees lose their leaves at the start of the dry season.

**dormant** describes a living thing that is resting before growing

**embryo** undeveloped plant contained in a seed

**energy** all living things need energy in order to live and grow and do everything—including breathing and eating—that they do

**evaporate/evaporation** when water turns from liquid into a vapor or gas. When clothes dry on a line, it is because the water evaporates and the vapor becomes part of the air.

**fat** nutrient that gives living things energy

**fertilization** when one flower's male sex cell joins with another flower's female sex cell and begins to form a seed

**flower** reproductive parts of a plant. Flowers contain the parts that can make seeds.

**fungi** (singular is *fungus*) plantlike living things that do not make their food by photosynthesis and that reproduce using spores. Mushrooms and molds are types of fungi.

**germinate/germination** when a seed starts to grow

**growth rings** rings you can see when the trunk of a tree is cut down. If you count the growth rings, you can find out how old a tree is.

**habitat**   place in the natural world where a particular organism lives

**inflorescence**   flower-like head that is made up of many tiny flowers growing closely together

**leaf**   plant part that makes the plant's food

**mangrove**   kind of shrub that grows in seashore mud or salt marshes

**midrib**   central vein that runs up through the middle of a leaf

**nectar**   sugary substance plants make to attract insects

**nectary**   place at the base of flower petals where nectar is made and stored

**nutrient**   chemical that plants and animals need in order to live

**organism**   living thing, such as a plant or animal

**ovary**   part of the flower that contains the female sex cells. It forms the bottom part of the pistil.

**petal**   colored part of a flower

**petiole**   stalk that attaches a leaf to a stem

**phloem**   tubes that carry sugars made in the leaf to all the other parts of the plant

**photosynthesis**   process by which plants make their own food using water, carbon dioxide, and energy from sunlight

**pistil**   name for the female parts of a flower. The ovary, style, and stigma together make up a pistil.

**pod**   capsule that holds the seeds of legume plants, such as peas

**pollen**   tiny dust-like particles that are produced by the anthers in the male part of a flower

**pollinate/pollination**   when pollen travels from the anthers of one flower to the stigma of the same or a different flower for the purpose of reproduction

**protein**   substance needed by living things for growth, maintenance, and repair of their bodies

**rain forest**   kind of forest habitat that exists in very rainy regions of the world

**receptacle**   swollen end of a flower nearest to the stem

**reproduce/reproduction**   when a living thing produces young like itself

**rhizome**   special kind of stem that grows under the ground

**root**   plant part that anchors the plant firmly in the ground and absorbs water and nutrients

**sap**   sugary liquid containing food made in the leaves. Sap flows in a plant's phloem tissue.

**seed**   part of a plant that contains the beginnings of a new plant

**seed leaf**   small leaf that stores food inside a seed. The embryo inside a seed uses the seed leaves for the energy it needs to grow.

**sepal**   green petal-like structure. Sepals protect the inner parts of a bud until the bud is ready to open.

**sex cell**  male or female cells that combine to make an embryo. Male sex cells are called sperm, and female sex cells are called ovules.

**shoot**  new stem growing from the main stem of a plant or out of a seed

**species**  kind of living thing. Organisms of the same species can breed together to produce fertile offspring.

**sporangia**  sacs, or capsules, containing spores

**stalk**  part of the plant that attaches the leaf to the stem. Flower stalks attach little flowers to a stem.

**stamen**  male pollen-producing reproductive part in a flower

**stem**  part of a plant that holds it upright and supports its leaves and flowers

**stomata**  (singular is *stoma*) tiny openings on a leaf, usually on the underside, that let water vapor and oxygen out and carbon dioxide in

**tendrils**  very long, thin leaves that wrap around supports to hold up a climbing plant

**tuber**  stem that grows underground and becomes swollen with food that the plant stores inside it. Potatoes are a kind of tuber.

**umbel**  group of small flowers that form an umbrella-like shape

**vein**  tiny vessel that helps support a leaf. Xylem and phloem vessels run through the veins, carrying water to the leaves and food made in the leaves to other parts of the plant.

**vitamin**  type of nutrient in food that helps animals grow and protects them from illness

**xylem**  tubelike tissue in a plant that carries water and nutrients from the roots to all the other parts of the plant

# Find Out More

## Books

Burnie, David. *The Plant*. New York: Dorling Kindersley, 2000.

Dorling Kindersley Publishing Staff. *Extraordinary Plants*. New York: Dorling Kindersley, 1997.

Joly, Dominique, et al. *How Does Your Garden Grow? Be Your Own Plant Expert*. New York: Sterling Publishing Company, 2000.

Madgwick, Wendy. *Flowering Plants: The Green World*. Collingdale, Penn.: DIANE Publishing Company, 2000.

Oxalade, Chris. *Flowering Plants*. Danbury, Conn.: Children's Press, 1999.

## Conservation sites

Center for Plant Conservation
P.O. Box 299
St. Louis, MO 63166-0299
Tel: (314) 577-9450

Lady Bird Johnson Wildflower Center
4801 La Crosse Avenue
Austin, TX 78739-1702
Tel: (512) 292-4200

New England Wild Flower Society
180 Hemenway Road
Framingham, MA 01701
Tel: (508) 877-7630

California Native Plant Society
1722 J Street, Suite 17
Sacramento, CA 95814
Tel: (415) 970-0394

The State Botanical Garden of Georgia
2450 S. Milledge Avenue
Athens, GA 30605
Tel: (706) 542-6448

## Places to visit

Many museums, arboretums (botanical garden devoted to trees) and botanic gardens are fascinating places to visit. You could try:

The New York Botanical Garden
Bronx River Parkway at Fordham Road
Bronx, NY 10458
Tel: (718) 817-8700

Denver Botanic Gardens
1005 York Street
Denver, CO 80206
Tel: (720) 865-3500

Atlanta Botanical Garden
1345 Piedmont Avenue NE
Atlanta, GA 30309
Tel: (404) 876-5859

Garfield Park Conservatory
300 North Central Park Ave.
Chicago, IL 60624-1996
Tel: (312) 746-5100

University of California Botanical Garden
200 Centennial Drive
Berkeley, CA 94720
Tel: (510) 643-2755

Missouri Botanical Garden
P.O. Box 299
St. Louis, MO 63166-0299
Tel: (314) 577-9400

# Index